WORKBOOK

11th Edition

Modern Residential Wiring

Nancy Henke-Konopasek

Harvey N. Holzman

Publisher
The Goodheart-Willcox Company, Inc.
Tinley Park, IL
www.g-w.com

Introduction

    This workbook is designed for use with the text *Modern Residential Wiring*. The chapters in the workbook correspond to those in the text and should be completed after reading the appropriate text chapter.

    Each chapter of the workbook reviews the material found in the textbook chapters to enhance your understanding of textbook content. The various types of questions include matching, true or false, multiple choice, fill-in-the-blank, and short answer.

    Reading *Modern Residential Wiring* and using this workbook will provide you with a solid background of electrical wiring principles and common practices, as well as a basic understanding of the *National Electric Code*®. Once this information is mastered, you will be prepared to further your knowledge of wiring methods through additional readings and practical experiences.

Contents

Name _____

Date _____

Class _____

CHAPTER **1**

Electrical Careers

Complete the following questions and problems after carefully reading the corresponding textbook chapter.

1. List seven electrical career categories.

   _____

   _____

   _____

   _____

   _____

   _____

   _____

_____ 2. *True or False?* One-fourth of the workers in the electrical power industry are involved in transmission and distribution jobs.

_____ 3. Line installers and repairers construct and _____ the network of power lines that carries electricity from generating plants to consumers.

_____ 4. Troubleshooters are experienced line installers and repairers who are assigned to special crews that handle _____.

_____ 5. When cables are installed, _____ are the people who pull the cable through the conduit and join the cables at connecting points.

_____ 6. _____ electricians follow blueprints and specifications for most installations.

_____ 7. Maintenance electricians keep _____ in good working order.
   A. lighting systems
   B. transformers
   C. generators
   D. All of the above.

8. Why must electrical inspectors be very familiar with the *NEC* and the local codes?

   _____

   _____

   _____

9. What are the three essential steps to take when you are ready to apply for jobs?

   _____

   _____

   _____

_____ 10. A brief outline of your education, work experience, and other
   qualifications for work is called a(n) _____.
   A. letter of application
   B. résumé
   C. application form
   D. job interview

11. List six items that should be included in a letter of application.

   _____

   _____

   _____

   _____

   _____

   _____

_____ 12. Which of the following is *not* a good tip to follow when completing
   a job application?
   A. Be as neat as possible.
   B. Write "open" or "negotiable" for any question regarding
      salary range.
   C. When asked about a former job, give negative comments.
   D. Complete every question on the form or write "NA" so the
      employer knows you did not overlook it.

Name _____

_____ 13. Which of the following is recommended when going on a job interview?
  A. Chew gum so that you appear relaxed.
  B. Dress appropriately, usually one step above what is worn by future coworkers.
  C. Arrive exactly on time.
  D. Use a weak handshake.

_____ 14. *True or False?* A punctual employee always starts the workday promptly and takes extended breaks and lunches.

_____ 15. Taking _____ means that you start activities on your own without being told.

_____ 16. Your _____ is your outlook on life and can often determine the success you have on your job.

_____ 17. Behaving professionally on the job includes which of the following?
  A. Showing respect for your boss.
  B. Acting courteously.
  C. Responding appropriately to constructive criticism.
  D. All of the above.

18. List the five steps in decision making and problem solving.

_____

_____

_____

_____

_____

_____ 19. Verbal communication involves all of the following *except* _____.
  A. speaking
  B. listening to what others say
  C. body language, such as facial expressions and posture
  D. writing

_____ 20. The most important aspect of customer relations is _____.
  A. letting the customer know how smart you are
  B. always remaining courteous
  C. making sure the customer knows you are an expert
  D. getting customers to purchase from your business no matter what

_____ 21. A big advantage of working in a(n) _____ is its ability to develop plans and complete work faster than individuals working alone.

_____ 22. The ability to guide and motivate others to complete tasks or achieve goals is _____.

_____ 23. Which of the following behaviors is *not* constructive in managing conflict?
    A. Making snap judgments.
    B. Treating others the way you would want to be treated.
    C. Exploring both positive and negative aspects of each possible solution.
    D. Trying to understand a problem from the other's point of view.

_____ 24. _____ is the process of agreeing to an issue that requires all parties to give and take.

_____ 25. *True or False?* A disadvantage of owning your own business is that the workdays are considerably longer for the "boss."

26. List six characteristics you should have if considering starting your own business.

_____

_____

_____

_____

_____

_____

_____ 27. A(n) _____ program may be sponsored by local unions or professional trade associations and includes both on-the-job training and related classroom instruction.

_____ 28. A(n) _____ is obtained by passing an examination that tests knowledge of electrical theory and its application, as well as national and local code requirements.

29. What is SkillsUSA?

_____

_____

_____

Name _____

30. List five organizations that can provide valuable information that may help you determine your future career path.

_____

_____

_____

_____

_____

NOTES

Name _____

Date _____

Class _____

CHAPTER **2**
Safety

Complete the following questions and problems after carefully reading the corresponding textbook chapter.

_____ 1. Using shortcuts to work around safe practices _____.
   A. saves you time
   B. saves you money
   C. can quickly become shortcuts to disaster
   D. is acceptable when you have confidence in doing the job

2. Complete the following chart.

Average Effects of Electric Current on the Body	
Amount of Current	**Effects on the Body**
0.001 A (1 mA)	
0.001 A to 0.01 A (1 mA to 10 mA)	
0.01 A to 0.1 A (10 mA to 100 mA)	
0.1 A or more (100 mA or more)	

Goodheart-Willcox Publisher

_____ 3. Death caused by electric shock is called _____.

4. Name and describe the two types of electricity-related burns that can occur.

_____

_____

_____

_____

5. Why is it important to know where you are working and be able to describe the work site location?

    _____

    _____

    _____

    _____

_____ 6. Always wear _____ whenever there is the slightest chance of an object being projected.

7. List four key safety goals to keep in mind.

    _____

    _____

    _____

    _____

8. List four pieces of personal protective equipment (PPE).

    _____

    _____

    _____

    _____

_____ 9. Ladders used by electricians should be made of _____.
    A. wood
    B. metal
    C. fiberglass
    D. Either A or C.

10. How far away should the base of a ladder be positioned from the structure?

    _____

    _____

_____ 11. *True or False?* It is a safe practice to climb the braces to get onto a scaffold.

Name _____

_____ 12. A tool is considered insulated if _____.
   A. it has a plastic handle
   B. it has a rubber handle
   C. the manufacturer certifies it for a specific level of electricity
   D. All of the above.

13. List six proper lifting procedures to reduce strain on your back.

_____

_____

_____

_____

_____

_____

_____ 14. To avoid contacting an underground conductor, _____.
   A. always notify the local utility company before digging
   B. do not use any powered excavating equipment within 24″ of the utility markings
   C. use a hand shovel within 24″ of utility markings
   D. All of the above.

_____ 15. Before you dig, call _____ to reach the national Call Before You Dig operator who will connect you to a local call center who will, in turn, contact all local utilities with lines in your area.

16. List three procedures that help prevent other workers from energizing a circuit being worked on by an electrician.

_____

_____

_____

_____ 17. *True or False?* Using insulated tools can help prevent shock in some situations, but should never be used as a substitute for turning off the power.

_____ 18. If a shock victim is still in contact with the source of electrical current, _____.
   A. move the conductor or victim with your hands
   B. use a wood stick or other nonconductive, insulated material to separate the conductor from the victim
   C. avoid shutting off the power
   D. wait for someone else to help

19. List three examples of common types of hazards found on construction sites.

_____

_____

_____

_____   20. Communication among crews working on the same job site is essential for workers' _____.

21. Once a hazardous situation has been recognized, reported, and repaired, you should review the situation by asking what five questions?

_____

_____

_____

_____

_____

22. Name five organizations that research, develop, and maintain rules, codes, and laws that are designed to promote safety.

_____

_____

_____

_____

_____

Name _____

Date _____

Class _____

CHAPTER **3**

Hand and Power Tools

Complete the following questions and problems after carefully reading the corresponding textbook chapter.

1. Identify the basic measuring tools pictured below.

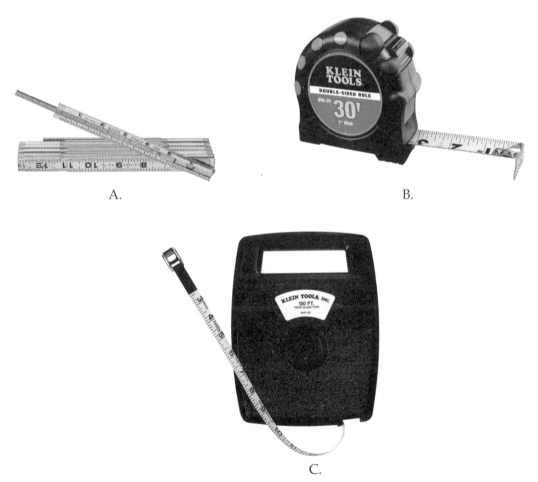

A.

B.

C.

Klein Tools, Inc.

A. _____

B. _____

C. _____

2. Identify the tool pictured below.

Klein Tools, Inc.

3. Identify the tools pictured below.

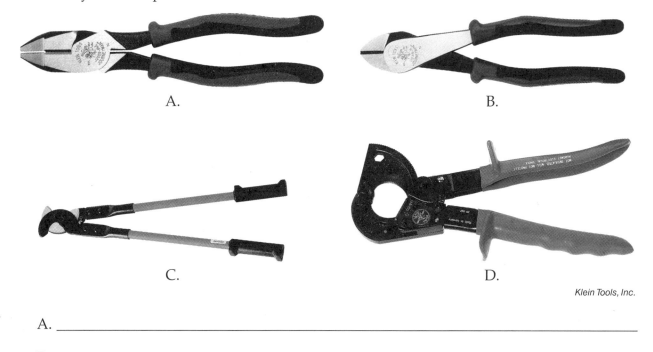

A.

B.

C.

D.

Klein Tools, Inc.

A. _____

B. _____

C. _____

D. _____

4. Identify the tool pictured below.

Klein Tools, Inc.

Name _____

5. Identify the tool pictured below.

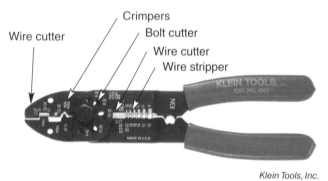

Klein Tools, Inc.

6. Identify the tool pictured below.

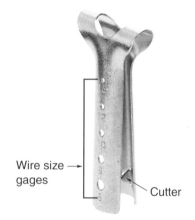

Klein Tools, Inc.

7. Identify the tool pictured below.

Klein Tools, Inc.

8. Identify the tool pictured below.

Klein Tools, Inc.

_____

9. Identify the types of pliers pictured below.

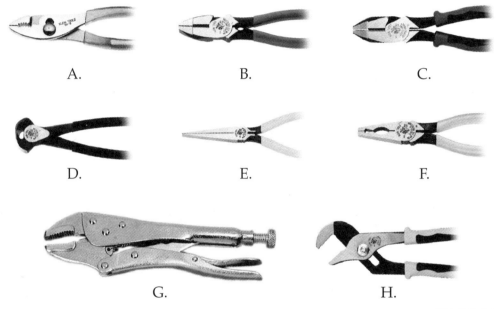

Klein Tools, Inc.

A. _____

B. _____

C. _____

D. _____

E. _____

F. _____

G. _____

H. _____

Name _____

10. Identify the tool pictured below.

dcwcreations/Shutterstock.com

11. Identify the tools pictured below.

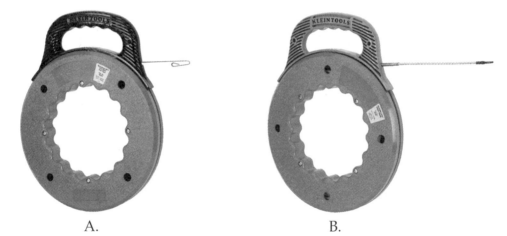

A.   B.

Klein Tools, Inc.

A. _____

B. _____

12. Identify the tools pictured below.

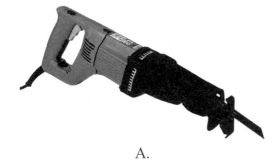 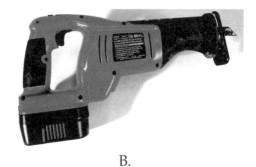

A.   B.

Makita Corp.

A. _____

B. _____

13. Identify the tool pictured below.

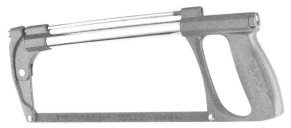

Klein Tools, Inc.

_____

14. Identify the types of vises pictured below.

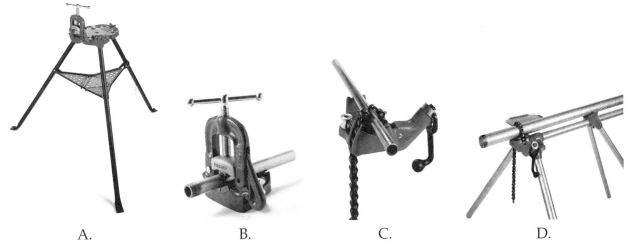

A.   B.   C.   D.

Ridge Tool Co.

A. _____

B. _____

C. _____

D. _____

15. Identify the tool pictured below.

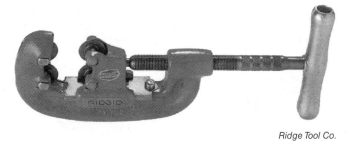

Ridge Tool Co.

_____

Name _____

16. Identify the type of tools pictured below.

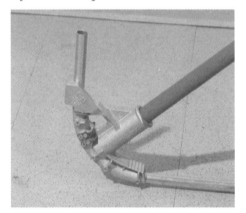

Goodheart-Willcox Publisher

17. Identify the tools pictured below.

Goodheart-Willcox Publisher   *Appleton Electric Co.*

18. Identify the types of hammers pictured below.

A. _____

B. _____

C. _____

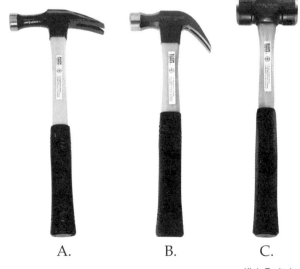

A.   B.   C.

Klein Tools, Inc.

19. Identify the drill tools pictured below.

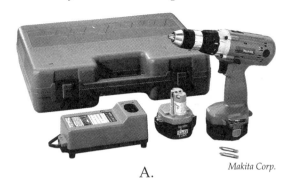

A.

Makita Corp.

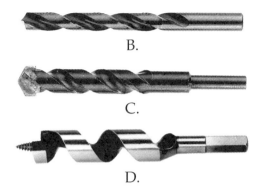

B.

C.

D.

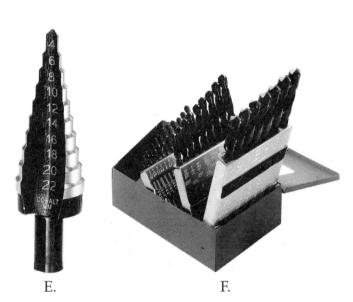

E.

F.

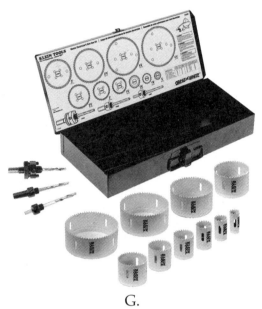

G.

Klein Tools, Inc.

A. _____

B. _____

C. _____

D. _____

E. _____

F. _____

G. _____

Name _____

20. Identify the tools pictured below.

A.

B.

Milwaukee Electric Tool Corp.

A. _____

B. _____

_____ 21. Knives, drill bits, chisels, saws, and other tools meant for cutting
should be kept _____.

_____ 22. If reconditioning a tool is impossible or costs more than a new
one, _____.
A. sell it
B. replace the tool
C. use the tool as is
D. All of the above.

_____ 23. Proper use and care of tools should be a part of every
electrician's _____.

24. Identify the following hand tools.

Goodheart-Willcox Publisher

A.

LifetimeStock/Shutterstock.com

B.

Goodheart-Willcox Publisher

C.

Klein Tools, Inc.

D.

Coleman Cable, Inc.

E.

Klein Tools, Inc.

F.

Klein Tools, Inc.

G.

A. _____

B. _____

C. _____

D. _____

E. _____

F. _____

G. _____

Name _____

25. Identify the following hand tools.

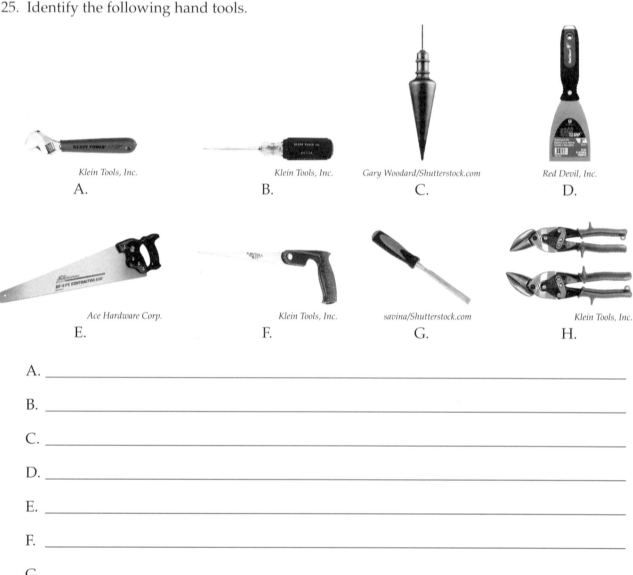

Klein Tools, Inc.
A.

Klein Tools, Inc.
B.

Gary Woodard/Shutterstock.com
C.

Red Devil, Inc.
D.

Ace Hardware Corp.
E.

Klein Tools, Inc.
F.

savina/Shutterstock.com
G.

Klein Tools, Inc.
H.

A. _____

B. _____

C. _____

D. _____

E. _____

F. _____

G. _____

H. _____

NOTES

Name _____

Date _____

Class _____

CHAPTER **4**

Electrical Measurement and Testing Equipment

Complete the following questions and problems after carefully reading the corresponding textbook chapter.

1. Why are meters and testers essential tools?

   _____

   _____

   _____

2. Identify the item pictured below and describe its use.

Ideal Industries, Inc.

   _____

   _____

   _____

3. Identify the item pictured below and describe its use.

Klein Tools, Inc.

_____

_____

_____

4. Identify the item pictured below and describe its use.

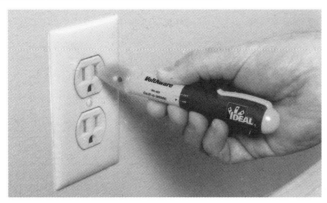

Ideal Industries, Inc.

_____

_____

_____

5. Identify the item pictured below and describe its use.

Ideal Industries, Inc.

_____

_____

_____

Name _____

_____ 6. *True or False?* A reliable way to double-check for safety is to use a tester on a circuit known to be live to make sure the tester itself is working before using it on the circuit on which you will be working.

7. Identify the items pictured below and describe how they are used.

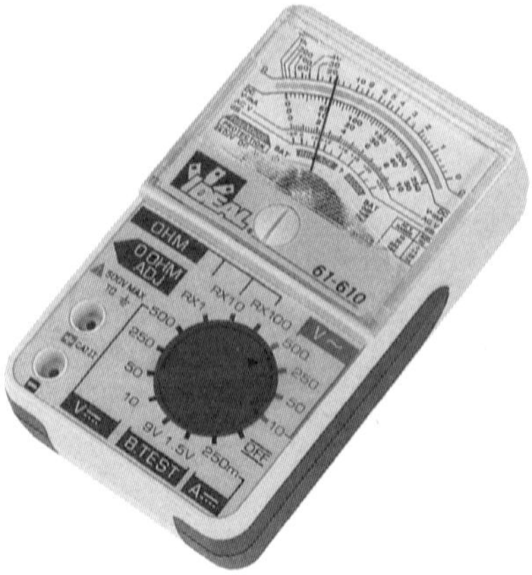

A.

Ideal Industries, Inc.

B.

Goodheart-Willcox Publisher

A. _____

B. _____

_____

_____

_____

8. List the six steps used to measure voltage at a receptacle using a multimeter.

_____

_____

_____

_____

_____

_____

_____

_____ 9. *True or False?* A multimeter can be damaged or destroyed if it is used to test a live circuit that has less current or voltage than the capability of the meter.

_____ 10. Multimeters are handy for testing _____, a simple test for determining whether or not a circuit is complete.

11. Identify the item pictured below and describe its use.

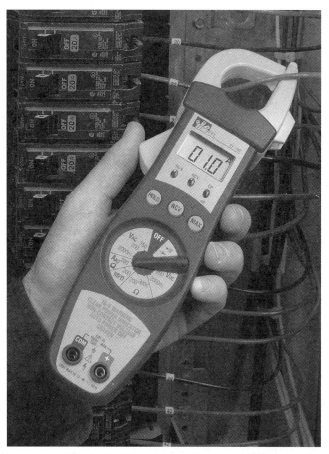

Ideal Industries, Inc.

_____

_____

_____

Name _____

12. Identify the item pictured below and describe its use.

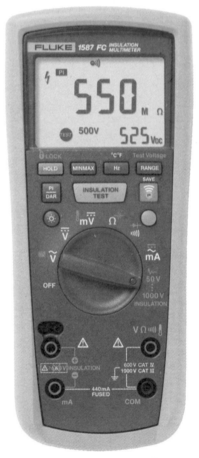

Reproduced with Permission, Fluke Corporation

_____

_____

_____

13. Identify the item pictured below and describe its use.

Reproduced with Permission, Fluke Corporation

_____

_____

_____

Name _____

14. Identify the item pictured below and describe its use.

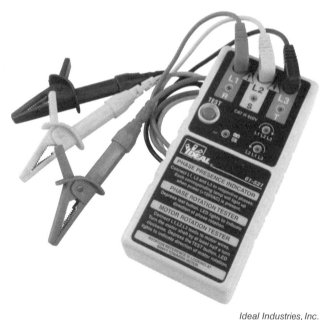

Ideal Industries, Inc.

_____

_____

_____

15. List eight basic guidelines for ensuring safety when measuring and testing.

_____

_____

_____

_____

_____

_____

_____

_____

NOTES

Name _____

Date _____

Class _____

CHAPTER **5**

Electrical Energy Fundamentals

Complete the following questions and problems after carefully reading the corresponding textbook chapter.

_____ 1. According to the _____ theory, all matter is made up of atoms.

Identify the components of an atom, as illustrated in the following figure.

_____ 2. Protons

_____ 3. Orbital path

_____ 4. Neutrons

_____ 5. Nucleus

_____ 6. Electrons

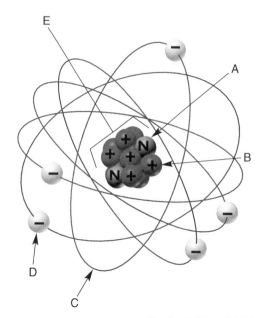

Goodheart-Willcox Publisher

_____ 7. Protons are _____ charged particles.

_____ 8. The _____ inside the nucleus of an atom have no charge.

_____ 9. Which of the following is an example of a good conductor?
   A. Glass
   B. Pure metal
   C. Dry gases
   D. All of the above.

_____ 10. Which of the following is an example of a good insulator?
   A. Most liquids
   B. Carbon
   C. Rubber
   D. All of the above.

_____ 11. Electrons will flow from a negatively charged body to a positively charged one. This flow is called _____.

_____ 12. In electricity, differences in potential are measured in _____.

13. List the two types of electric current.

_____

_____

_____ 14. In _____ current, the electricity flows in only one direction.

_____ 15. In the United States, _____ current provided by the power company has a frequency of 60 Hz.

_____ 16. A(n) _____ is one complete revolution of a generator.

_____ 17. A(n) _____ is a group of voltaic cells that provide electrical power.

_____ 18. A wet cell battery contains multiple cells made up of metal plates suspended in a(n) _____.
   A. cathode
   B. anode
   C. electrolyte
   D. None of the above.

_____ 19. In a battery, the pole that is positively charged is the _____.

_____ 20. In a battery, the pole that is negatively charged is the _____.

_____ 21. A generator is driven by some _____ force.

_____ 22. Which of the following units is used to measure electricity?
   A. Amperage
   B. Voltage
   C. Wattage
   D. All of the above.

_____ 23. The unit used to measure current is the _____.

24. What are two names for the electrical pressure or force by which electrons are moved through a conductor?

_____

_____

Name _____

_____ 25. The units for measuring electrical pressure are _____.

_____ 26. The opposition to the flow of electrons through a conductor is called _____.

_____ 27. Resistance is measured in _____.

28. What are the two units used to measure power?

_____

_____

_____ 29. Work is measured in units called _____.

_____ 30. A kilowatt-hour is equivalent to _____ being used for one hour.
A. 100 W
B. 1,000 W
C. 1,000,000 W
D. None of the above.

_____ 31. In electrical wiring, the pathway for electron flow is called a(n) _____.

32. A simple circuit consists of what four elements?

_____

_____

_____

_____

33. List the three types of electrical circuits.

_____

_____

_____

_____ 34. A(n) _____ circuit has only one path for the current.

_____ 35. A(n) _____ circuit has more than one path for the current.

_____ 36. What type of circuit is represented in the following diagram?

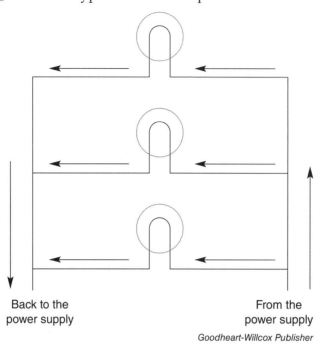

Back to the
power supply

From the
power supply

Goodheart-Willcox Publisher

_____ 37. What type of circuit is represented in the following diagram?

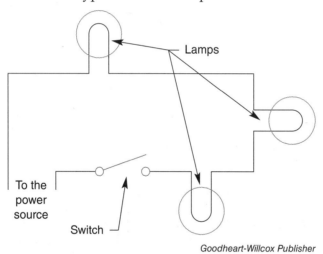

Lamps

To the
power
source

Switch

Goodheart-Willcox Publisher

_____ 38. *True or False?* Series circuitry is practical for residential wiring.

_____ 39. In electrical wiring, a(n) _____ is any device that uses an electric current and converts the energy to another form.

Name _____

40. Define *Ohm's law*.

_____

_____

_____

_____

_____

41. The following diagram presents a simple way to aid remembering Ohm's law. Describe how to use it.

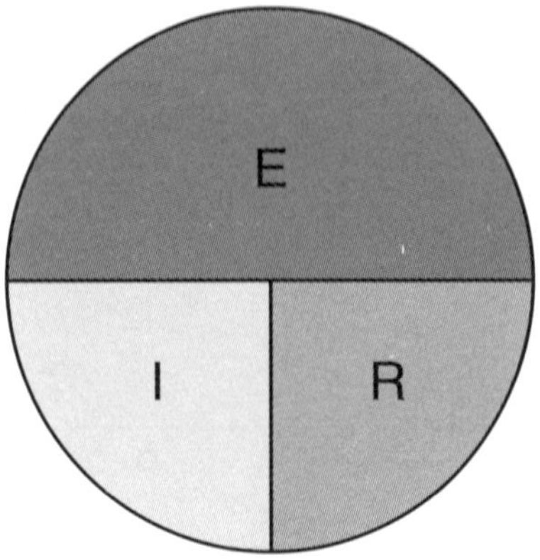

Goodheart-Willcox Publisher

_____

_____

_____

_____

_____

_____ 42. *True or False?* Electrical power is current times voltage.

_____ 43. The rate at which an electrical circuit consumes energy is _____.

44. How much power is a circuit using if it is drawing 7 A of current and has a total voltage of 12 V?

_____

_____

45. List five rules that apply to series circuits.

_____

_____

_____

_____

_____

_____

_____

Use the following series circuit diagram to answer the next two questions.

$R_3 = 30$ ohms

$E_2 = 30$ volts
$R_2 = 10$ ohms

$R_1 = 20$ ohms

Goodheart-Willcox Publisher

_____ 46. What is the total resistance in this series circuit?

_____ 47. What is the current in this series circuit?

_____ 48. In electricity, the term *parallel* means _____.
   A. physically parallel
   B. geometrically parallel
   C. alternate routes
   D. None of the above.

49. List five rules that apply to parallel circuits.

_____

_____

_____

_____

_____

_____

_____

Name _____

50. What is the total current of a parallel circuit with a total resistance of 15 Ω and a source voltage of 30 V?

_____

_____

_____ 51. A(n) _____ circuit is a combination of a series circuit and a parallel circuit.

52. What is electromagnetic induction? How is it accomplished?

_____

_____

_____

_____

_____

_____ 53. _____ is at its highest level when a conductor is moving at a right angle to magnetic field lines of force.

Match the parts of an alternating current generator with the corresponding description.

_____ 54. Coil or armature

_____ 55. Nonmoving poles or opposite ends of field magnets

_____ 56. Slip rings

_____ 57. Brushes

A. These are connected to the ends of the coil wires.

B. This part rotates. It has the conducting wire that cuts across a magnetic field.

C. Two of these transfer current from the slip rings to the external circuit.

D. These create the magnetic field.

_____ 58. A dc generator is similar in construction to an alternator except that the two slip rings are replaced by one slip ring with two halves, which is commonly called a(n) _____.

_____ 59. *True or False?* An ac generator may produce either single-phase or three-phase electric power.

_____ 60. For high-demand, industrial applications, _____ -phase power is used.

_____ 61. *True or False?* Three-phase motors are simpler, less expensive, and more powerful than single-phase motors.

_____ 62. A(n) _____ is a device that uses electromagnetic induction to change the voltage as it transfers electrical energy from one circuit to another.

_____ 63. Voltage out of a transformer is directly related to the number of _____ of wire in the windings.

_____ 64. In a step-down transformer, the number of turns of wire is greater in the primary winding than in the secondary winding, so the output voltage will be _____ the input voltage.
A. less than
B. greater than
C. the same as
D. None of the above.

_____ 65. Before electric power is brought into households, it is reduced to _____.
A. 120 V and 240 V
B. 132,000 V
C. 300 V and 600 V
D. 750,000 V

_____ 66. The power enters a house through the _____, which is where the electric meter is located.

_____ 67. _____ will shut off power in a circuit if the circuit becomes overloaded.

CHAPTER **6**

Print Reading, Specifications, and Codes

Complete the following questions and problems after carefully reading the corresponding textbook chapter.

_____ 1. A(n) _____ is a copy of a drawing that illustrates the materials, locations, and methods that must be used during construction.

_____ 2. Which of the following shows the major land-based features?
A. Framing prints
B. Foundation prints
C. Plot prints
D. None of the above.

_____ 3. Which of the following illustrates features ranging from the overall shape of a house to the small details of trimming out a window?
A. Foundation prints
B. Plot prints
C. Framing prints
D. None of the above.

_____ 4. Symbols that represent lighting fixtures, outlet receptacles, switches, panels, and the circuits that connect them are shown on the _____ print.

_____ 5. The _____ on a print shows the drawing name, the names of the people who worked on the drawing, and the drawing number and scale.

_____ 6. You can find changes and corrections made on a drawing listed in the _____.

_____ 7. *True or False?* Most electrical prints are 17″ × 22″ or larger.

_____ 8. The _____ of a drawing is the ratio of the size of the object in real life to the size that the object was reduced or enlarged to in the drawing.

Standard electrical symbols should be understood and memorized by every electrician. Commonly used symbols that an electrician will need to know are given below. Identify each of the following symbols.

General Outlets

9. ○   ─○ _____

10. Ⓔ ─Ⓔ _____

11. Ⓕ ─Ⓕ _____

12. Ⓛ ─Ⓛ _____

13. Ⓛ$_{PS}$ ─Ⓛ$_{PS}$ _____

14. Ⓒ ─Ⓒ _____

Receptacles

15. ⊨⊖ _____

16. ⊨⊖$_3$ _____

17. ⊨⊖$_{WP}$ _____

18. ⊨⊖ _____

19. ⊨⊖$_R$ _____

20. ⊨⊖$_S$ _____

21. ⊨⊖$_{GFCI}$ _____

22. ⊨⊖ _____

23. ◓ _____

24. ⊟ _____

Name _____

Switches

25. S   _____

26. S_2 _____

27. S_3 _____

28. S_4 _____

29. S_K _____

30. S_P _____

31. S_{CB} _____

32. S_{WCB} _____

33. S_{MC} _____

34. S_{WP} _____

35. S_T _____

Miscellaneous

36. ▬   _____

37. ▬   _____

38. ⓣ   _____

39. ⊏▭⊐   _____

40. ⊏OR⊐   _____

41. ◁   _____

42. ⊣◁   _____

_____ 43. *True or False?* A special outlet symbol with a lowercase letter beside it must be listed in a drawing's legend.

44. Explain why the lines representing circuits on electrical prints are always curved.

_____

_____

_____

45. List five common types of electrical schedules that may be needed for a large project.

_____

_____

_____

_____ 46. *True or False?* Local notes contain information that applies to an entire drawing.

47. Why are specifications included with the prints?

_____

_____

_____

_____

_____

_____ 48. The _____ establishes a set of rules, regulations, and criteria for the installation of electrical equipment.

_____ 49. *True or False?* Although the *NEC* has no legal basis, it is often made mandatory under local or state rulings. In such cases, it becomes a legal document.

50. What is the purpose of a nationally recognized testing laboratory (NRTL)?

_____

_____

_____

_____

_____

CHAPTER **7**

Branch-Circuit, Feeder, and Service Design

Complete the following questions and problems after carefully reading the corresponding textbook chapter.

_____ 1. *True or False?* Each branch circuit represents a load that contributes to the total load for the household system.

_____ 2. A(n) _____ is a separate electrical path, independent of other electrical paths in the building.

3. Explain why it is practical to use branch circuits when wiring a building.

_____

_____

_____

_____ 4. The _____ is a percentage of the total load on a circuit that is subtracted from the total load to create a lower electrical load.

_____ 5. The total of all branch circuit loads can be calculated after you know the _____.
   A. square footage of living space
   B. number and size of any large appliances
   C. largest size of any noncoincident loads
   D. All of the above.

_____ 6. The *Code* specifies that _____ volt-amperes per square foot of living space should be used to calculate the general lighting load.

7. Explain how to calculate the general lighting load of a home.

_____

_____

_____

_____ 8. *True or False?* When calculating the number of square feet in a house to find the total general lighting load, include the garage in the total if it is not adaptable for future use as living space.

_____ 9. The load for a specific appliance is based on the ampere rating specified on the _____ of the appliance.

_____ 10. _____ loads are loads that will not be operated at the same time.

11. What is the purpose of an overcurrent protection device (OCPD)?

_____

_____

_____

_____ 12. *True or False?* A subpanel and feeder increase the length of the home run for every branch circuit fed by the subpanel.

13. Identify two situations that can be prevented by having the proper number of branch circuits.

_____

_____

_____

14. List five *NEC* requirements for bathrooms.

_____

_____

_____

_____

_____

_____

_____

_____ 15. Kitchens must have a minimum of two _____-amp circuits dedicated to receptacles for small appliances.

_____ 16. *True or False?* Laundry equipment uses a 20-ampere circuit that has many other receptacles.

Name _____

_____ 17. Lights should be controlled by a _____.
  A. switch near each entrance to a room
  B. pull chain on the fixture itself
  C. pushbutton on the fixture itself
  D. All of the above.

_____ 18. *True or False?* A light switch should be located on the latch side of a doorway.

_____ 19. Receptacle outlets must be located so that no point along any wall space is more than _____ from a receptacle.

_____ 20. Receptacle outlets should be placed 12 to 15 inches above the _____.

21. List 10 places where receptacle outlets are required in a residence. Also, indicate where GFCI receptacles should be used.

_____

_____

_____

_____

_____

_____

_____

_____

_____

_____

_____ 22. *True or False?* Every room should have at least one wall-switched lighting outlet.

List a basic residential lighting requirement for each room indicated in the next eight questions.

23. Rooms

_____

_____

24. Stairs

_____

_____

25. Basement

_____

_____

26. Kitchen and bathrooms

_____

_____

27. Garage

_____

_____

28. Entrances

_____

_____

29. Hallways

_____

_____

30. Attic

_____

_____

_____ 31. When planning the hookup of branch circuits to the service panel, it is important to keep a(n) _____ in the load between the two hot wires in a three-wire system.

Name _____

Date _____

Class _____

CHAPTER **8**

Conductors

Complete the following questions and problems after carefully reading the corresponding textbook chapter.

_____ 1. The term _____ is used to represent only the metal portion of a wire that conducts electricity.

_____ 2. A(n) _____ is a metal conductor and the insulation that surrounds it.

Identify the items indicated on the illustration of insulated wires that follows.

_____ 3. Conductor

_____ 4. Stranded wire

_____ 5. Insulation

_____ 6. Solid wire

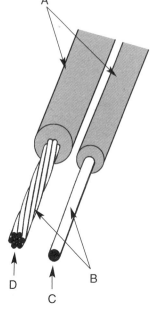

Goodheart-Willcox Publisher

7. List three considerations that must be made when choosing the right wire or cable.

    _____

    _____

    _____

_____ 8. Only _____ and _____ conductors are used in residential wiring.

_____ 9. Which of the following is the most common conductor material
                                for electrical wiring due to its strength and resistance to oxidation?
                                A. Aluminum
                                B. Copper
                                C. Rubber
                                D. Plastic

_____ 10. Which of the following conductor materials is subject to problems
                                as a result of oxidation and expansion?
                                A. Tin
                                B. Rubber
                                C. Aluminum
                                D. Copper

_____ 11. The most common insulation material used in electrical wiring is a
                                thermoplastic made from _____.

_____ 12. Type _____ is the most common wire used in dry installations,
                                while _____ is frequently used in wet installations.

13. List three factors that determine the size of the conductor to be used for electrical wiring.

    _____

    _____

    _____

14. What is the American Wire Gage?

    _____

    _____

_____ 15. In the AWG, the _____ (larger, smaller) the wire number, the
                                smaller the diameter.

_____ 16. A(n) _____ is the area of a circle with a diameter of 0.001″ and is
                                the unit used to measure the cross-sectional area of wires.

Name ______________________________

17. List two important factors affected by wire size.

____

____

18. Stripping tools designed for solid wire cannot be used on stranded wire. Why?

____

____

____________________ 19. *True or False?* Temperature has no effect on ampacity and does not need to be considered during installation.

____________________ 20. The ampacity of a conductor _____ as the temperature increases.
   A. increases
   B. decreases
   C. stays the same

____________________ 21. A(n) _____ is a percentage used to reduce the ampacity of a conductor based on the number of current-carrying conductors in a raceway or cable.

For the next eight questions, identify what information is indicated by the markings on the cable covering shown below.

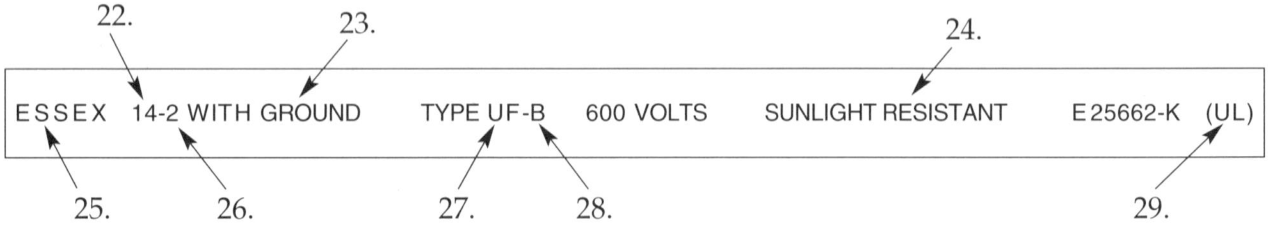
22.
23.
24.
ESSEX 14-2 WITH GROUND TYPE UF-B 600 VOLTS SUNLIGHT RESISTANT E25662-K (UL)
25.
26.
27.
28.
29.

Goodheart-Willcox Publisher

22. ____

23. ____

24. ____

25. ____

26. ____

27. ____

29. ____

____________________ 30. Color-coding of _____ indicates the purpose of the conductor.

NOTES

Name _____

Date _____

Class _____

CHAPTER **9**

Cable Systems

Complete the following questions and problems after carefully reading the corresponding textbook chapter.

_____ 1. A(n) _____ is an arrangement of two or more conductors in a protective covering of plastic, rubber, steel, or aluminum.

_____ 2. Flexible metal cable is _____.
A. used in situations when the cable is exposed
B. used where the additional protection of a metal jacket is required
C. sold without wires inside
D. Both A and B.

_____ 3. _____ is a manufactured assembly of conductors in a flexible interlocked metallic armor.

_____ 4. *True or False?* Armored cable is often used in commercial garages, elevators, and theaters.

5. List three tools that can be used to cut through armored cable.

_____

_____

_____

_____ 6. A(n) _____ is a red, plastic sleeve that prevents the wire insulation from rubbing against the sharp edge of the armor.

_____ 7. *True or False?* Metal-clad cable contains a ground wire.

_____ 8. _____ cable is a flexible cable that is used for most residential circuits and is typically installed without conduit.

_____ 9. *True or False?* Nonmetallic sheathed cable is the easiest and, in many areas, the most popular system to install.

_____ 10. Which type of nonmetallic sheathed cable may be used only in dry
locations?
A. NM
B. NMC
C. UF
D. USE

11. Identify the NM cable fittings shown in the following illustrations.

NM Cable Fittings

A.

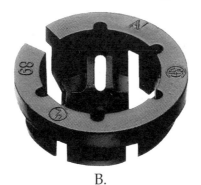
B.

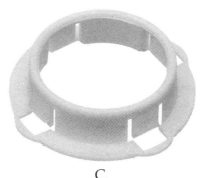
C.

Goodheart-Willcox Publisher

A. _____

B. _____

C. _____

_____ 12. _____ cable is used to bring electricity from the utility company's
point of attachment to the customer's equipment.

_____ 13. A typical installation of underground feeder and branch-circuit
cable would be a branch from the main service panel to a _____.
A. transformer
B. detached garage
C. swimming pool pump
D. All of the above.

_____ 14. UF cable is resistant to _____.
A. sunlight
B. moisture
C. fungus
D. All of the above.

Name _____

Date _____

Class _____

CHAPTER **10**

Raceway Systems

Complete the following questions and problems after carefully reading the corresponding textbook chapter.

_____ 1. _____ is the tubing or channel that is used to connect boxes together.

_____ 2. *True or False?* Cutting EMT and other types of conduit creates a rough edge that must be removed to prevent damage to the wire's insulation.

_____ 3. The *NEC* specifies the _____ of conduit bends.

4. List four typical bends.

_____

_____

5. Explain when an offset bend would be used.

_____

_____

_____

_____

_____ 6. If an offset bend is going to occur before the obstruction, the amount of _____ must be considered.

7.   What is the difference between a saddle bend and an offset bend?

_____

_____

_____

_____

_____

_____

_____   8. *True or False?* The most common saddle bend consists of two 45° center bends and one 22 1/2° lateral bend.

_____   9. Electrical nonmetallic tubing (ENT) is _____.
   A. lightweight and flexible
   B. for voltage exceeding 600 V
   C. used for supportive means
   D. used for direct burial applications

_____   10. _____ is galvanized and somewhat similar to water pipe.

_____   11. Which of the following conductor systems is cut, threaded, and reamed before it is installed?
   A. EMT or thin-wall conduit
   B. Nonmetallic sheathed cable
   C. Rigid metal conduit
   D. Flexible metal conduit

_____   12. *True or False?* Intermediate metal conduit (IMC) is permitted for use in all atmospheric conditions and in all types of occupancies.

_____   13. Rigid polyvinyl chloride conduit (PVC) _____.
   A. can be used to support fixtures
   B. can be used in hazardous locations where physical damage is likely
   C. can be used with various types of fittings, elbows, and mounting hardware
   D. All of the above.

_____   14. *True or False?* Rigid polyvinyl chloride conduit (PVC) can be bent by hand when it is warm enough.

_____   15. *True or False?* Flexible metal conduit, or Greenfield, cannot be used with EMT or rigid conduit.

Name _____

16. Explain why Greenfield is frequently used with EMT or rigid conduit.

_____

_____

_____

_____

_____ 17. Because of its extreme flexibility, _____ can be used to connect machinery that is portable or may vibrate during normal operation.

_____ 18. *True or False?* Surface-mount raceway is useful in existing facilities when appearance is important and new wiring cannot be pulled through walls.

NOTES

CHAPTER **11**
Boxes, Fittings, and Covers

Complete the following questions and problems after carefully reading the corresponding textbook chapter.

_____ 1. _____ are used to mount devices, conduit, and cables, as well as providing an enclosure for junctions.

_____ 2. The *NEC* requires which of the following to be housed inside approved enclosures?
A. Joints
B. Connections
C. Splices
D. All of the above.

3. Name three materials from which electrical boxes may be made.

_____

_____

_____

_____

_____

For the next four questions, identify the four common box shapes used for electrical wiring boxes.

A

B

C

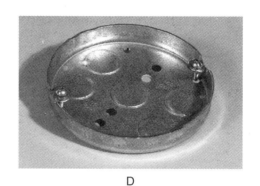

D

Goodheart-Willcox Publisher

_____ 4. Rectangle

_____ 5. Square

_____ 6. Octagonal

_____ 7. Round

_____ 8. A ceiling box may be _____.
　　　　　　　A. square
　　　　　　　B. round
　　　　　　　C. octagonal
　　　　　　　D. All of the above.

_____ 9. Wall boxes are usually _____ in shape.
　　　　　　　A. square
　　　　　　　B. round
　　　　　　　C. octagonal
　　　　　　　D. rectangular

_____ 10. *True or False?* Almost any type of shape of box is suitable as a pull box.

_____ 11. _____ boxes are seamless and have rounded corners to prevent injuries.

_____ 12. A(n) _____ is a pre-cut or pre-scored hole on the box that is designed to be easily removed so wires can be brought into the box.

Name _____

_____ 13. _____ is a method of joining small boxes to create one large box that can accommodate multiple devices.

_____ 14. *True or False?* Boxes must be securely fastened to a structural member of the wall, ceiling, or floor of the dwelling.

_____ 15. Nonmetallic boxes are _____.
A. inexpensive
B. lightweight
C. corrosion-resistant
D. All of the above.

_____ 16. _____ boxes are required for installations using metal conduit, armored cable (AC) or MC cable, and also can be used with NM cable.

_____ 17. *True or False?* An old work box has special features that allow it to be installed after the drywall is up.

_____ 18. _____ are parts of a wiring system designed to interconnect conduit, conductors, or boxes.

Identify the fittings for electrical wiring shown in the illustrations that follow.

_____ 19.

_____ 20.

_____ 21.

_____ 22.

_____ 23.

_____ 24.

Electroline Fittings LLC

_____ 25.

_____ 26.

_____ 27.

_____ 28.

_____ 29.

_____ 30.

_____ 31.

_____ 32.

_____ 33.

_____ 34.

Electroline Fittings LLC

Name _____

_____ 35.

_____ 36.

Electroline Fittings LLC

37. Identify three purposes of box extension rings.

_____

_____

_____

_____

38. List two fittings that secure cables and conduit to an electrical box.

_____

_____

_____ 39. _____ are connected to conduit to create a wire pull point.

_____ 40. *True or False?* Grounding is required only on metal boxes.

_____ 41. A(n) _____ is a spring device that is hooked over the edge of the box and holds the wire in tight electrical contact with the box.

_____ 42. The *NEC* requires a(n) _____ wherever conduit enters a nonthreaded opening on a box.

_____ 43. _____ refers to the number of conductors that the *NEC* will allow in certain sizes of boxes.

_____ 44. *True or False?* Plaster rings compensate for the thickness of the drywall, plaster, or other finish material.

_____ 45. Receptacles serving outdoor areas must have special _____ covers designed for the application.

NOTES

Name _____

Date _____

Class _____

CHAPTER 12

Device Wiring

Complete the following questions and problems after carefully reading the corresponding textbook chapter.

_____ 1. Proper mechanical installation of electrical equipment means _____.
   A. equipment is securely mounted
   B. conductors and cables are carefully routed, supported, and terminated to prevent damage
   C. conductors are properly secured to their termination points
   D. All of the above.

_____ 2. *True or False?* All electrical connections must be made inside an electrical box or enclosure.

_____ 3. All splices and connections must be covered with insulation _____ (less than, equal to) the conductor's original insulation.

_____ 4. To ensure accessibility and to maintain, service, or operate electrical equipment, the *NEC* insists the work clearances shall not be less than _____ inches wide in front of equipment.

5. Explain why panelboard circuit directories must be correctly labeled and located at the panelboard.

_____

_____

_____

_____

_____ 6. In removing insulation from a conductor, you must be careful not to damage either the _____ or the remaining insulation.

_____   7. As a general rule, the amount of conductor that should be bared to make a proper connection is _____.
     A. a little less than 1″
     B. at least 5″
     C. 1′
     D. None of the above.

_____   8. The preferred way to remove insulation is with a(n) _____.

_____   9. When attaching a conductor to a device terminal, the curved hook on the conductor must be connected _____.
     A. so that it has less than a two-thirds wrap
     B. counterclockwise onto the terminal
     C. clockwise onto the terminal
     D. so that it overlaps

Identify the method of attaching a conductor to a receptacle illustrated in the following four questions.

10.

Slater Electric Inc.

11.

Slater Electric Inc.

Name _____

12.

Slater Electric Inc.

13.

Slater Electric Inc.

_____ 14. _____ is the process of connecting one wire to one or more wires.

15. List four types of connectors that can be used to join heavy wires, such as 6 AWG or larger.

_____

_____

_____

_____ 16. The function of a(n) _____ is to control the flow of electricity to one or more electrical devices.

_____ 17. *True or False?* A single-pole switch has four terminals.

_____ 18. A(n) _____ is a device used to transfer electrical energy from conductors to plug-in electrical equipment.

19. Identify the switches pictured below.

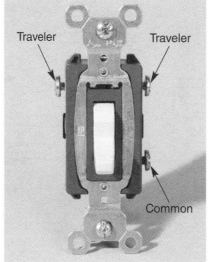

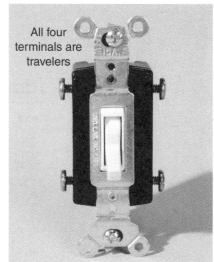

A.   B.   C.

Goodheart-Willcox Publisher

A. _____

B. _____

C. _____

_____ 20. The _____ is a removable tab that electrically connects the two outlets on a duplex receptacle.

_____ 21. In basic receptacle wiring, _____ wires connect to the silver or light-colored terminals, _____ conductors connect to the brass or dark-colored screws, and the _____ conductor connects to the ground screw.
   A. ground, neutral, hot
   B. neutral, hot, ground
   C. hot, neutral, ground
   D. ground, hot, neutral

Name _____

22. What method of wiring multiple receptacles on the same circuit is shown in the illustration below?

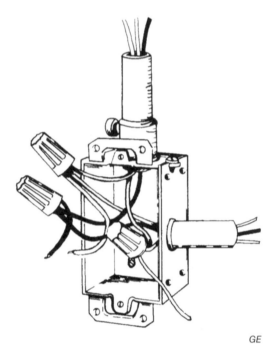

GE

_____

_____ 23. A(n)_____ receptacle is a duplex receptacle wired so that one of the two outlets is controlled by a switch while the other is always energized.

_____ 24. _____ boxes divide a circuit into several directions.

_____ 25. *True or False?* The grounding wire is shown in pictorial drawings for metallic conduit.

NOTES

Name _____

Date _____

Class _____

CHAPTER **13**

Lighting Systems

Complete the following questions and problems after carefully reading the corresponding textbook chapter.

_____ 1. The lighting industry and the *NEC* use the term _____ to describe a light source, such as a lightbulb, that produces visible light.

_____ 2. The *NEC* defines a(n) _____ as a complete lighting unit consisting of a light source, such as a lamp or lamps, together with the parts designed to position the light source and connect it to the power supply.

3. List the four categories of lighting effects or uses.

_____

_____

_____

_____ 4. General lighting _____.
   A. is the primary type of lighting in a home
   B. supplies broad, diffuse light that covers a large area
   C. is characterized as a comfortable brightness without glare and can be thought of as the light that simply fills the room
   D. All of the above.

_____ 5. _____ lighting provides strong, focused light to illuminate a work area.

_____ 6. _____ lighting is used to create visual interest or drama.

_____ 7. Security fixtures should be located _____.
   A. between 10′ and 12′ above the ground
   B. between 5′ and 6′ above the ground
   C. to the side of the detection area or target area
   D. Both A and C.

8. List the two basic types of ceiling fixtures.

_____

_____

_____ 9. *True or False?* Generally, hanging fixtures are less practical and more decorative than standard ceiling fixtures.

_____ 10. _____ fixtures consist of multiple movable fixtures connected to an energized track or other mounting system.

11. List one advantage and one disadvantage of recessed lighting.

Advantage: _____

_____

Disadvantage: _____

_____

_____ 12. A fixture in an insulated ceiling should carry a(n) _____ rating, indicating that it can be covered with thermal insulation.

_____ 13. _____ recessed fixtures have airtight housings or other features that prevent conditioned room air from escaping into the attic.

_____ 14. *True or False?* Wall fixtures are not desirable for lighting of a bathroom mirror.

_____ 15. *True or False?* A low-voltage fixture is not necessarily energy efficient.

16. Identify the different types of lamps that have Edison bases shown below.

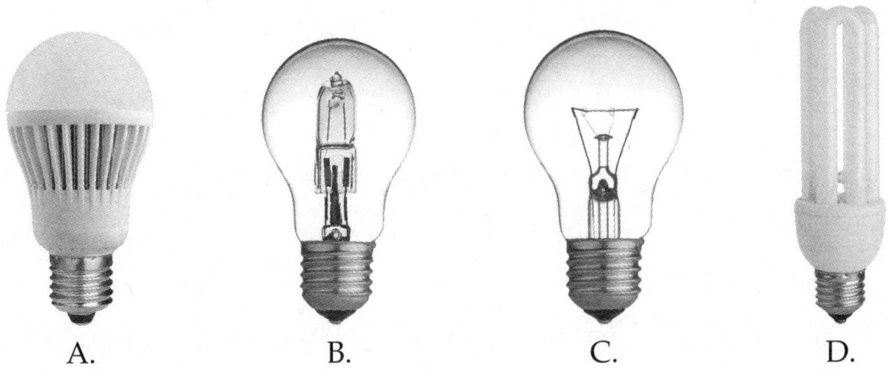

A.   B.   C.   D.

Chones/Shutterstock.com

A. _____

B. _____

C. _____

D. _____

Name _____

_____ 17. Up to _____% of the energy used by an incandescent lamp is wasted.
        A. 20
        B. 60
        C. 90
        D. 100

18. Explain why halogen lamps should not be changed with bare hands.

_____

_____

_____

_____

19. What might happen if a higher wattage lamp is used in a fixture rated for 60 watts?

_____

_____

_____

_____

_____ 20. _____ lamps produce light by conducting an electric current through an inert gas.

_____ 21. *True or False?* CFLs are four or five times more efficient than incandescent lamps.

_____ 22. LED lamps _____.
        A. do not contain mercury
        B. perform well in cold outdoor temperatures
        C. do not have a waiting time to reach full brightness
        D. All of the above.

_____ 23. When evaluating lamps for brightness, look at the number of _____ for an accurate comparison.

24. List the three factors to consider when comparing lamp costs.

_____

_____

_____

25. List five types of special switches.

_____

_____

_____

_____ 26. *True or False?* While standard switches and receptacles typically have screw terminals for making wire connections, most light fixtures have wire leads for connecting to circuit conductors.

27. List four factors that determine how fixtures should be mounted.

_____

_____

_____

_____

28. Describe what is shown in the illustration below.

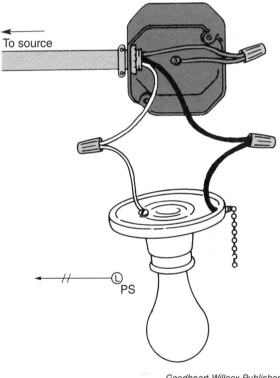

Goodheart-Willcox Publisher

_____

_____

_____

_____

Name _____

29. Describe what is shown in the illustration below.

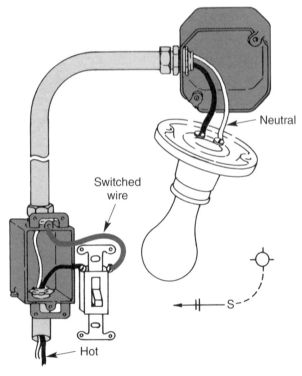

Neutral

Switched wire

Hot

Goodheart-Willcox Publisher

_____

_____

_____

_____

_____

_____

_____

30. Describe what is shown in the illustration below.

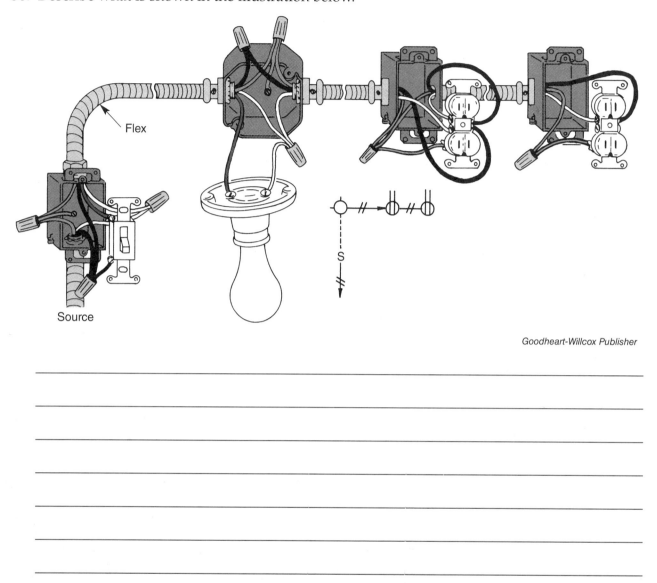

Goodheart-Willcox Publisher

_____

_____

_____

_____

_____

_____

_____

_____

Name _____

31. Describe what is shown in the illustration below.

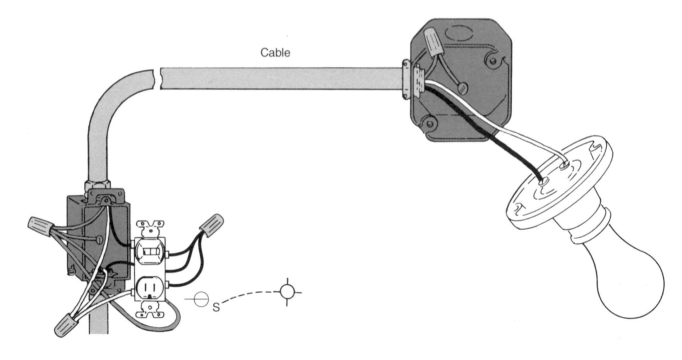

Cable

_____

_____

_____

_____

_____

_____

_____

32. Describe what is shown in the illustration below.

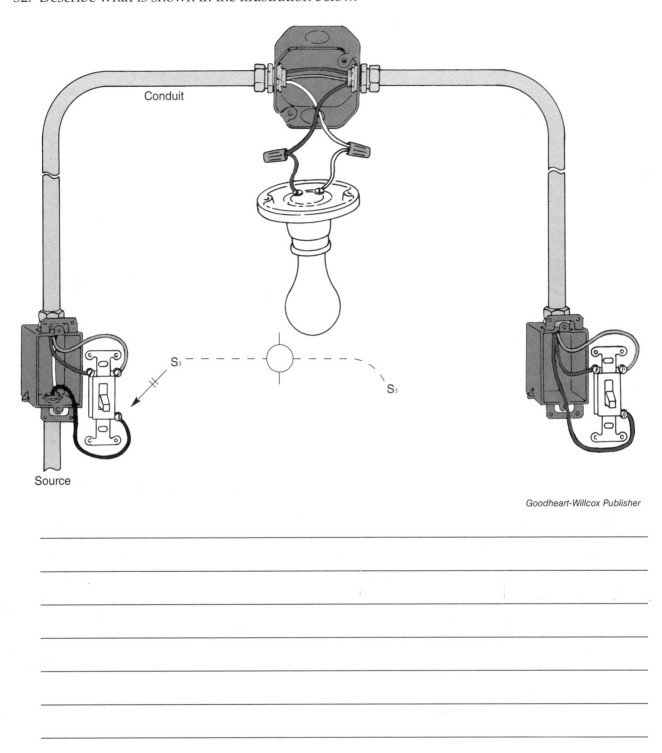

Conduit

Source

S₃

S₃

_____

_____

_____

_____

_____

_____

_____

Name _____

33. Describe what is shown in the illustration below.

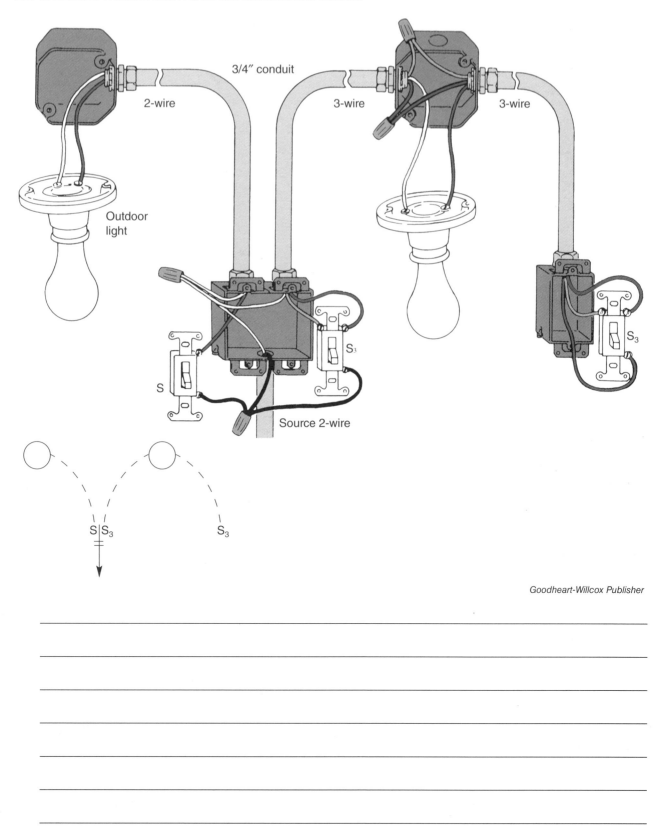

3/4" conduit

2-wire

3-wire

3-wire

Outdoor light

Source 2-wire

S

S₃

S₃

S | S₃   S₃

_____
_____
_____
_____
_____
_____
_____

34. Describe what is shown in the illustration below.

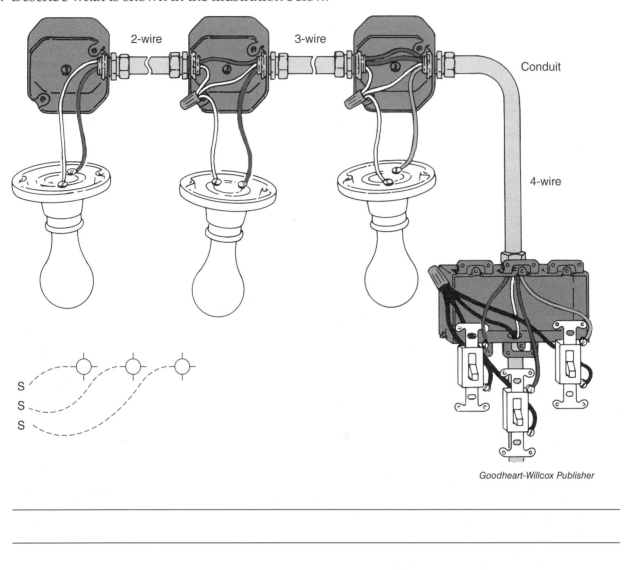

Goodheart-Willcox Publisher

_____

_____

_____

_____

Name _____

Date _____

Class _____

CHAPTER **14**

Appliance Wiring and Special Outlets

Complete the following questions and problems after carefully reading the corresponding textbook chapter.

1. List five large appliances that require special wiring considerations.

   _____

   _____

   _____

_____ 2. Hookups for clothes dryers and electric ranges are commonly made with a(n) _____ and a receptacle.

_____ 3. In general, appliances should be placed on a separate circuit if they are rated at or above _____.
   A. 12 A
   B. 1/8 horsepower
   C. 240 V
   D. All of the above.

_____ 4. Garbage disposal units are most often controlled _____.
   A. by a plug and cord connection
   B. from the panel circuit breaker
   C. by an on-off switch
   D. None of the above.

_____ 5. *True or False?* Most dishwashers carry a rating within the range of 5–10 A at 120 V.

_____ 6. Refrigerators and freezers should have overcurrent protection, fuses, or breakers rated at not more than _____ percent of their nameplate current.

_____ 7. Counter cooking tops and wall-mounted cooking units require special circuits and overcurrent protection in the form of fuses or breakers, usually _____ or _____ amps.

_____ 8. _____ can often be connected to the same circuit as heating units because they would not operate at the same time.

_____ 9. Electric water heaters require a separate _____-volt circuit.

10. Identify and define the three methods by which heat travels.

_____

_____

_____

_____

_____

_____

_____ 11. Heaters are almost always automatically controlled by _____ that are located close to the heating unit or built into it.

_____ 12. *True or False?* The *NEC* specifies that heaters should have a nameplate that states the size of the heater either in volts and watts or in volts and amperes.

13. List six factors that can influence the choice of heater units.

_____

_____

_____

_____

_____

_____

Name _____

14. Identify the parts of a baseboard heater circuit shown in the following diagram.

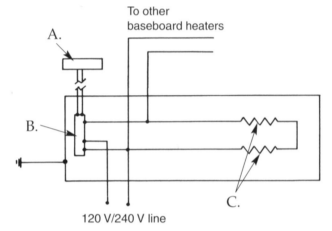

Goodheart-Willcox Publisher

A. _____

B. _____

C. _____

NOTES

Name _____

Date _____

Class _____

CHAPTER **15**

Overcurrent Protection

Complete the following questions and problems after carefully reading the corresponding textbook chapter.

_____ 1. A(n) _____ shuts off power to a circuit when the current in that circuit exceeds the ampacity rating of a specified amount.

_____ 2. Overcurrent may be a result of a(n) _____.
  A. circuit overload
  B. ground fault
  C. short circuit
  D. All of the above.

_____ 3. A(n) _____ is a situation where equipment operates beyond the rated ampacity of that equipment or the conductors within the circuit.

_____ 4. A(n) _____ is an overcurrent caused by the unintended connection of an ungrounded conductor to a grounded conductor.

_____ 5. A(n) _____ occurs when an ungrounded conductor touches a path to ground, such as a grounded raceway, a box, a fitting, or an equipment-grounding conductor.

_____ 6. A(n) _____ is a high-power discharge of an electric arc caused by current flowing through an uninterrupted path.

7. List two OCPDs.

_____

_____

_____ 8. *True or False?* The rating of an OCPD must not be greater than the overall capacity of the circuit being protected.

9. List the two general types of fuses.

_____

_____

_____ 10. *True or False?* Plug fuses with a circular window have a lower
rating than plug fuses with a hexagonal window.

_____ 11. *True or False?* It is safe to use a 30-amp Edison-base plug fuse in a
15-amp circuit.

12. Identify the types of cartridge fuses shown in the following illustration.

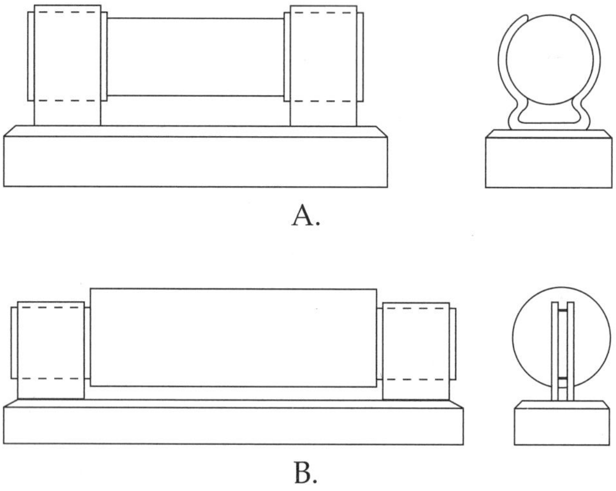

A.

B.

Bussmann Division of Cooper Industries

A. _____

B. _____

_____ 13. Which fuse has a low interrupting rating that restricts its use in
new construction?
A. Class H
B. Class J
C. Class R
D. Class T

_____ 14. Which fuse may be used as replacements for Class H fuses?
A. Class J
B. Class R
C. Class T
D. None of the above.

Name _____

15. List the steps that should be taken before installing a new fuse.

_____

_____

_____

_____

_____

_____ 16. A(n) _____ can open a circuit manually or automatically, and can be reset after it trips.

_____ 17. A single-pole breaker serves a _____-volt circuit.

_____ 18. A double-pole breaker serves a _____-volt circuit.

_____ 19. The _____ is the maximum amount of current with which an overcurrent protection device will reliably function.

_____ 20. A(n) _____ is a device that opens a circuit if an unusually large current-to-ground is detected.

21. List eight locations in which the *NEC* requires GFCI protection.

_____

_____

_____

_____

_____

_____ 22. A(n) _____ detects arcing faults that are not detected by standard breakers or GFCIs.

NOTES

Name _____

Date _____

Class _____

CHAPTER **16**

Grounding

Complete the following questions and problems after carefully reading the corresponding textbook chapter.

1. List two kinds of grounds for electrical wiring.

    _____

    _____

_____ 2. System grounding is the intentional connection of one _____ of an electrical system to the earth.

3. Explain why grounding electrical systems is required.

    _____

    _____

    _____

    _____

    _____

    _____

Identify the various parts indicated on the following schematic illustration of both system and equipment grounding.

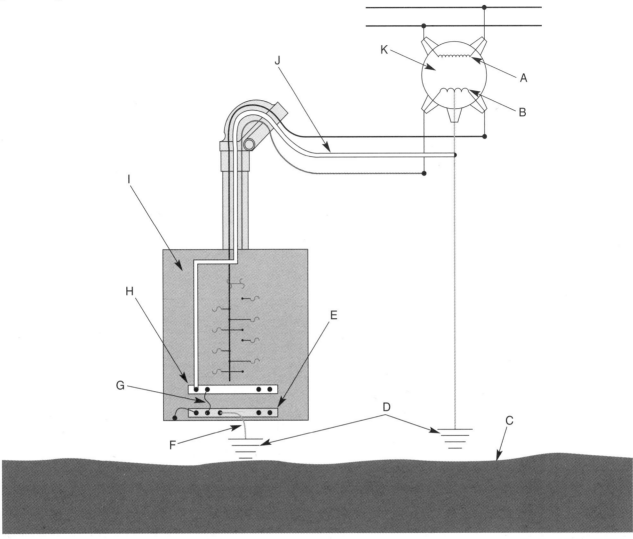

_____ 4. Grounding electrode

_____ 5. Primary

_____ 6. Neutral bus

_____ 7. Earth

_____ 8. Bonding jumpers

_____ 9. Secondary

_____ 10. Transformer

_____ 11. Grounding conductor

_____ 12. Main service panel

_____ 13. Grounding electrode conductor

_____ 14. Grounding bus

Name _____

_____ 15. *True or False?* A grounded conductor should never have its continuity interrupted by switches, fuses, or circuit breakers.

16. The size of the GEC is based on what three factors?

_____

_____

_____

_____ 17. Another method of system grounding is to connect the ground bus to part of the metal piping of the building's _____.

_____ 18. Joining all metal parts of the wiring system—boxes, cabinets, enclosures, and conduit—to ensure good, continuous metallic connections throughout the grounding system is called _____.

_____ 19. Bonding is required at all _____.
    A. conduit connections at the electrical service equipment
    B. service equipment enclosures
    C. metallic components of the electrical system that normally carry no current
    D. All of the above.

_____ 20. _____ grounding is the connecting of the equipment grounding conductor to equipment enclosures and metallic noncurrent-carrying equipment.

_____ 21. A(n) _____ occurs when conductor insulation fails or when a wire comes loose from its terminal point and touches the equipment housing or frame.

NOTES

Name _____

Date _____

Class _____

CHAPTER **17**

Service Equipment

Complete the following questions and problems after carefully reading the corresponding textbook chapter.

_____ 1. All electrical energy supplied to power-consuming devices in a building must first pass through the _____.

2. List the electrical equipment and parts of the service.

_____

_____

_____

_____ 3. When service conductors are routed underground, they are called the _____.

_____ 4. When conductors go to a building from overhead, they are called the _____.

5. List seven guidelines an electrician should follow when choosing the location of a service entrance.

_____

_____

_____

_____

_____

_____

_____

_____

_____ 6. *True or False?* In most cases, a structure can have only one service.

_____ 7. For new, single-family dwellings, the *NEC* requires at least a(n) _____-amp service.

_____ 8. Most service drops in new homes are made with _____ cable.

Identify the items indicated on the following service entrance diagram.

_____ 9. Roof

_____ 10. Insulators

_____ 11. Rain boot

_____ 12. Conduit

_____ 13. Mast

_____ 14. Service entrance conductors

_____ 15. Stud

_____ 16. Meter socket

_____ 17. Pipe clamp with lag screw into structure

_____ 18. Watertight hub

_____ 19. Service head

_____ 20. Siding

_____ 21. Sheathing

Insulator is mounted on the mast

Goodheart-Willcox Publisher

_____ 22. A(n) _____ is a fitting installed at the top of the service mast or service entrance cable to prevent water from entering the meter socket and shorting out the conductors.

_____ 23. The _____ is the device that measures and records the amount of electricity used.

_____ 24. *True or False?* A weatherproof connector is used to connect service entrance cable to the meter enclosure when conduit is not used.

Name _____

25. List five situations in which temporary wiring would be used.

_____

_____

_____

_____ 26. The *NEC* requires that disconnect means or service entrance conductors cannot exceed _____ switches either in a single enclosure or a group of separate enclosures.
   A. three
   B. five
   C. six
   D. eight

_____ 27. *True or False?* The disconnect must be located in a readily accessible place that is as close as possible to the point where the service conductors enter the structure.

28. Why is a grounding electrode conductor typically made of copper, aluminum, or copper-clad aluminum?

_____

_____

29. On the following figures, indicate the above-roof clearance for overhead conductors.

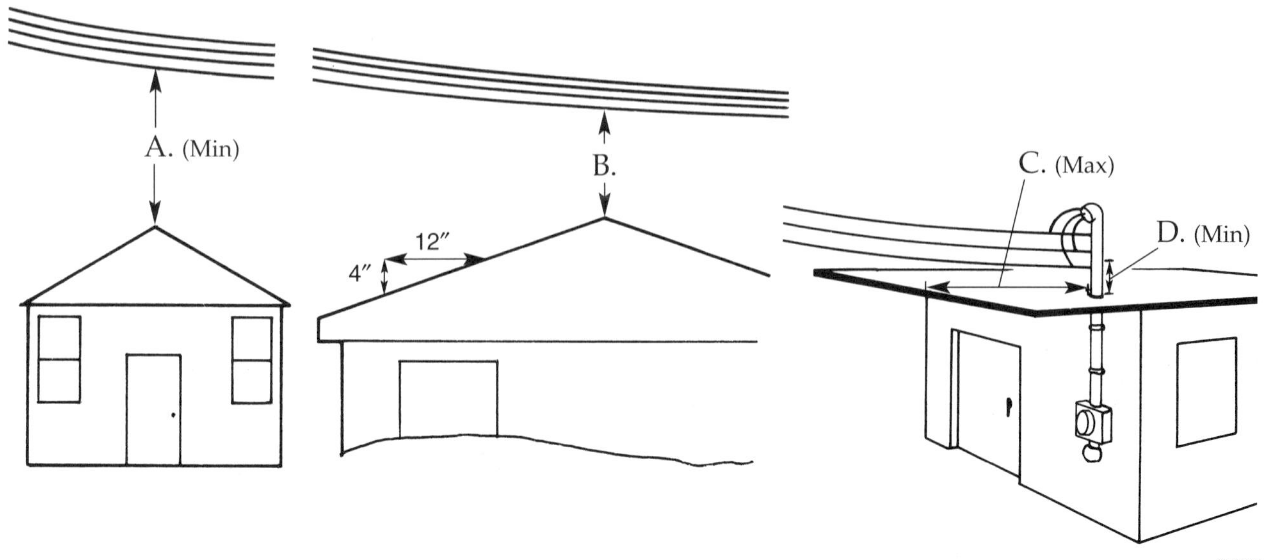

OSHA

A. _____   C. _____

B. _____   D. _____

30. On the following figure, indicate the above-grade minimum clearances.

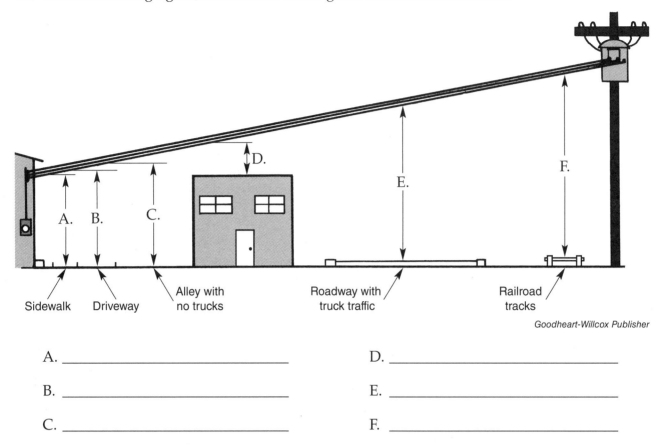

Sidewalk   Driveway   Alley with no trucks   Roadway with truck traffic   Railroad tracks

Goodheart-Willcox Publisher

A. _____   D. _____

B. _____   E. _____

C. _____   F. _____

31. On the following figures, indicate the clearances around building platforms and openings.

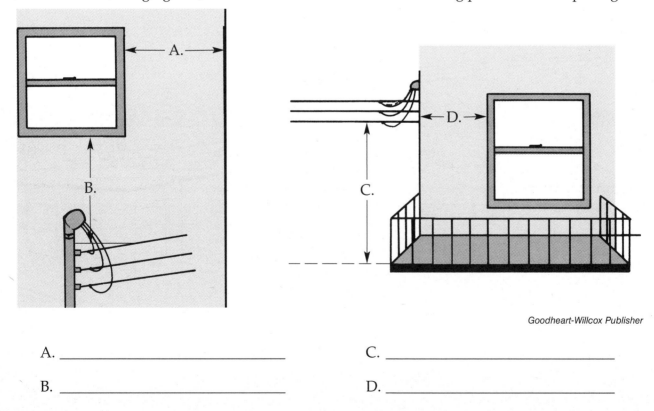

Goodheart-Willcox Publisher

A. _____   C. _____

B. _____   D. _____

Name _____

Identify the items indicated on the following diagram.

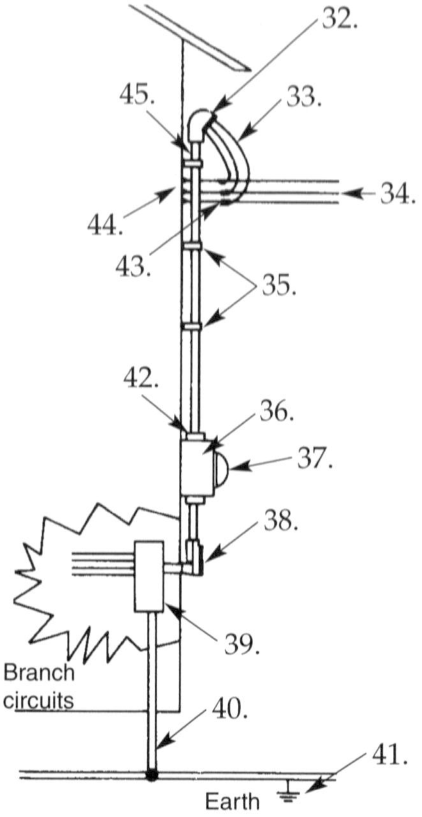

_____ 32.

_____ 33.

_____ 34.

_____ 35.

_____ 36.

_____ 37.

_____ 38.

_____ 39.

_____ 40.

_____ 41.

_____ 42.

_____ 43.

_____ 44.

_____ 45.

Goodheart-Willcox Publisher

_____ 46. *True or False?* Transformers that increase primary voltage are called step-down transformers.

_____ 47. *True or False?* A chain of transformers links the generating plant with the consumer.

48. What type of service is the most common residential service? How much voltage does it provide?

_____

_____

_____

_____

_____ 49. Three-phase, four-wire, _____-connected service supplies 120 volts of single-phase, 240 volts of single-phase, and 240 volts of three-phase circuits.

_____ 50. The phase B wire of the delta, four-wire, three-phase system, often called the "high leg" or the "wild leg," must be identified as such at all accessible points by a(n) _____-colored indicator.
A. green
B. purple
C. orange
D. white

Name _____

Date _____

Class _____

CHAPTER **18**

Farm Wiring

Complete the following questions and problems after carefully reading the corresponding textbook chapter.

_____ 1. A(n) _____ should be thought of as a small industrial plant or business that has several buildings.

_____ 2. At a farm, the service drop ends at a centrally located _____.

_____ 3. Almost all modern farms require a(n) _____ A to _____ A service.

_____ 4. *True or False?* On a farm, the yard pole serves as a power center and overall disconnecting means for the farm.

_____ 5. *True or False?* To ensure proper grounding and safe wiring in farm buildings, metal equipment is preferred.

_____ 6. A(n) _____ light fixture is good for the damp and dusty conditions on a farm.

_____ 7. In farm buildings, receptacles should be mounted no lower than _____ inches above the floor.

_____ 8. If ceilings are dark, use _____ so light is directed downward to where it is needed.

_____ 9. In a poultry house, _____ switches, which are used to increase egg production, may be a part of the lighting circuits.

_____ 10. *True or False?* The farmhouse power requirements are calculated the same way as they are for any residence.

_____ 11. *True or False?* The farm building that will probably have the greatest electrical demand is the barn.

_____ 12. Power _____ of electric motors of different sizes must be known in order to calculate a motor's load demand on an electrical circuit.

_____ 13. The ____ load represents the power that would be needed if everything were operating at the same time.

_____ 14. The ____ load represents the amount of power that would most probably be needed at any given time.

_____ 15. When considering future growth of a farm, it is recommended that sizes of service conductors be at least ____ size(s) larger than calculated.

Name _____

Date _____

Class _____

CHAPTER **19**

Mobile Home Wiring

Complete the following questions and problems after carefully reading the corresponding textbook chapter.

_____ 1. The _____ corresponds to the electrical service panel in a permanent structure.

_____ 2. The _____ includes the conductors, with their fittings, that carry electrical current from the mobile home service equipment to the distribution panelboard inside the mobile home.

_____ 3. A single power cord with a molded plug meets *NEC* requirements for a mobile home when the electrical load does not exceed _____ amperes.

4. List three ways the electrical feed can be supplied to a mobile home.

_____

_____

_____

_____ 5. The _____ is all of the equipment for connecting the supply end of a mobile home feeder assembly.

_____ 6. *True or False?* Service equipment must be mounted on the mobile home itself.

_____ 7. An overhead feeder hookup must have _____ continuous conductors.

_____ 8. *True or False?* For a raceway installation, the mobile home may be fitted with a flexible raceway of any material.

_____ 9. *True or False?* The manufacturer of the mobile home typically supplies the mobile home distribution panelboard.

_____ 10. *True or False?* In the distribution panelboard, the neutral conductor is completely insulated and isolated from the grounding conductor and the equipment enclosure.

_____ 11. *True or False?* Calculating the panelboard load for mobile homes is similar to calculating the panelboard load for standard residences.

12. If a mobile home park needs to accommodate 45 mobile homes (16,000 VA per mobile home lot) with 240 V service, what would the amperage be for the park's main service entrance? (The demand factor is 23 percent.)

_____

_____

_____ 13. *True or False?* Common mobile home park facilities and equipment, such as security lighting, must be considered in the total service demand.

Name _____

Date _____

Class _____

CHAPTER **20**

Swimming Pool Wiring

Complete the following questions and problems after carefully reading the corresponding textbook chapter.

1. What are seven types of swimming pool components that must be bonded with an 8 AWG solid copper conductor?

    _____

    _____

    _____

    _____

    _____

    _____

    _____

_____   2. The only job of the bonding conductor is to _____ the metal parts.

_____   3. In addition to bonding, all electrical equipment related to the pool and within 5′ of the interior pool walls must be properly _____.

_____   4. Receptacles used for _____ motors and the circulation and sanitation systems must be at least 6′ (1.83 m) away from the inside walls of the pool.

_____   5. All switches or switching devices of any type must be set back at least _____ from the pool edge.

6. List the eight categories of pool lighting.

    _____

    _____

    _____

    _____

    _____

    _____

    _____

    _____

_____ 7. *True or False?* The *NEC* requires that existing lighting fixtures be removed if they are within the prohibited 5′ perimeter of the poolside.

_____ 8. A(n) _____ luminaire is a light fixture intended for installation in a forming shell that is mounted in a pool or fountain where the fixture will be completely surrounded by water.

_____ 9. A(n) _____ luminaire is a light fixture intended for installation in the floor or wall of a pool, spa, or fountain in a niche that is sealed against entry of water.

_____ 10. *True or False?* For all electrical items to be wired within or around a swimming pool, care should be taken to ensure proper grounding and bonding as a number one priority.

_____ 11. Overhead utility conductors must not be within _____ horizontally of the inside wall of the pool.

Name _____

Date _____

Class _____

CHAPTER 21

Telephone and Computer Network Wiring

Complete the following questions and problems after carefully reading the corresponding textbook chapter.

_____ 1. *True or False?* Some systems have discarded the traditional phone system entirely, opting to carry voice signals over the Internet.

_____ 2. The _____ is the telephone connection from the building to the central office.

_____ 3. The _____ is made up of two wires that carry the signals and power to a telephone.

_____ 4. The enclosure where connections are made for the cable pair going to each building is called the _____.

_____ 5. A(n) _____ is a device that prevents high voltage from damaging the cable pair should lightning strike the wires.

_____ 6. The junction point for all the telephones in a building is called the _____.

_____ 7. Computer _____ allow PCs to share files, printers, and an Internet access device.

_____ 8. Keep telephone and network wiring at least _____ inches away from power circuits in a building.

_____ 9. Jacks should be kept away from _____.
    A. showers and tubs
    B. swimming pools
    C. laundry areas
    D. All of the above.

10. List the three types of indoor telephone cable.

_____

_____

_____

_____ 11. _____ is the phenomenon of a current carrying conductor imposing an electrical signal on another conductor.

12. What are the two general types of twisted-pair cables?

_____

_____

13. List the three categories of twisted-pair cable that are typically used for residential networks.

_____

_____

_____ 14. A(n) _____ is a length of twisted-pair cable that is terminated on each end with an RJ-45 plug.

15. List five rules that should be followed when running cables.

_____

_____

_____

_____

_____

_____

_____

_____

_____

_____

_____ 16. *True or False?* Electrical metallic tubing (EMT) is used to guard against electromagnetic interference.

_____ 17. *True or False?* In new construction, the telephone wiring should be installed after the application of drywall or paneling.

Name _____

Identify the items indicated on the faceplate image shown below.

_____ 18.

_____ 19.

_____ 20.

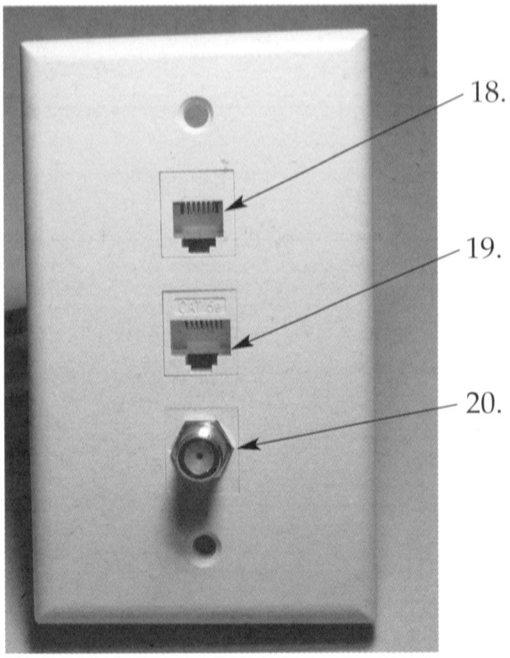

Goodheart-Willcox Publisher

_____ 21. Telephone wires can be made less noticeable by placing them _____.
   A. in areas where moldings or grooves in paneling partially or wholly conceal them
   B. inside cabinets
   C. under carpet
   D. Both A and B.

_____ 22. A(n) _____ is a long length of cable that is typically pulled through walls and the ceiling area.

_____ 23. The standard computer plugs are _____.

_____ 24. Fiber optic cable is a clear glass or plastic conductor that transmits _____ signals.

_____ 25. _____ cable contains coaxial, twisted pair, and sometimes fiber optic cables into one bundle.

NOTES

Name _____

Date _____

Class _____

CHAPTER **22**

Motors

Complete the following questions and problems after carefully reading the corresponding textbook chapter.

1. List five types of motors.

    _____

    _____

    _____

    _____

    _____

_____ 2. The _____ provides a wealth of data regarding the characteristics of a motor.

3. List the information that the *NEC* and the NEMA require to appear on the nameplates of all motors.

    _____

    _____

    _____

    _____

    _____

    _____

_____ 4. A(n) _____ rating indicates the lengths of time a loaded motor can run in an hour or how frequently a motor can be turned on or off.

_____   5. The _____ assigned to each integral horsepower motor ensures the shaft height and other physical dimensions are consistent.

_____   6. If applied voltage varies too much from the nameplate specifications, it will produce noticeable changes in the motor _____.

_____   7. Voltage drop greatly affects the _____ and _____ of a motor.

8. List four common motor circuits.

_____

_____

_____

_____

_____   9. The overload protection in the starter must have a rating that is not more than _____ percent of the motor's full-load amperage.

_____   10. _____ fuses are designed mainly for motor-circuit protection.

11. List four requirements that a single motor branch circuit must provide.

_____

_____

_____

_____

_____

_____

_____   12. When each motor on a circuit is 1 hp or 6 A or less, the circuit will require a(n) _____ A/125-volt or smaller fuse or breaker.

_____   13. In a motor branch circuit, the overcurrent protection must not exceed the amperage stamped on the OCPD of the _____ (smallest, largest) motor on the branch circuit.

_____   14. *True or False?* A single disconnect for a group of motors is never permitted.

Name _____

15. What is a motor controller? What is its purpose?

_____

_____

_____

_____

_____

_____ 16. The OCPD may serve as the controller for motors that are rated not more than _____ horsepower, stationary, and normally left running.

_____ 17. A switch may be the controller for motors rated up to and including _____ horsepower.

_____ 18. A cord and plug arrangement may serve as the controller for motors at or less than _____ horsepower.

_____ 19. _____ motors can handle heavier loads than general-purpose motors because they stay cool during normal operation.

_____ 20. _____ selection for hermetic motors is based on the full-load amperage and locked-rotor amperage values.

21. Define *rated-load current*.

_____

_____

_____

_____

_____ 22. The disconnecting-means rating for a hermetic motor shall be 115 percent of the nameplate rated-load current or the branch circuit selection current, whichever is _____ (less, greater).

23. List five things you must know in order to properly install a circuit with a combination load.

_____

_____

_____

_____

_____

24. List four common causes of motor failure.

_____

_____

_____

25. List four factors that can cause a motor to overheat.

_____

_____

_____

_____

_____ 26. A(n) _____ is a hole in a motor or other enclosure that is designed to allow any moisture to drain out and prevent accumulation of moisture.

_____ 27. _____ may seize in unused motors that are not rotated for extended periods.

_____ 28. Secure mounting and correct _____ with the load are essential for proper motor performance.

29. List three ways motors may be connected to a load.

_____

_____

_____

_____ 30. *True or False?* A time-delay relay should be used to restart motors randomly after a power loss in order to prevent an excessive voltage drop from all the motors restarting at one time.

Name _____

Date _____

Class _____

CHAPTER **23**

Emergency and Standby Systems

Complete the following questions and problems after carefully reading the corresponding textbook chapter.

1. Explain why you should never fill a gasoline tank indoors.

   _____

   _____

   _____

   _____

_____ 2. _____ is a colorless, odorless, and tasteless gas that can accumulate in unventilated areas and can cause headaches, nausea, and death.

_____ 3. A(n) _____ connects the home's circuit either to the service conductors from the utility or to the generator.

_____ 4. A cord that has a male plug on each end is called a(n) _____ and should never be used.

5. What is the purpose of emergency systems?

   _____

   _____

   _____

_____ 6. *True or False?* A portable generator is a low-cost power supply that can generate enough power to operate every appliance in a home at once.

_____ 7. A simple double-pole double-throw switch can be used as a(n) _____ switch.

_____ 8. A permanent generator with an automatic transfer switch _____.
  A. is wired into the main circuit breaker
  B. eliminates having to plug in a cord and throw a switch
  C. can be programmed to start once a week and run for a short period to keep the engine lubricated
  D. All of the above.

9. What is the first step in installing a generator?

_____

_____

_____ 10. *True or False?* The automatic transfer switch should be located near the main service panel.

_____ 11. The only way to ensure a constant flow of electricity is to use a(n) _____.

_____ 12. *True or False?* The transition from the main power supply to the battery in the UPS is so instantaneous that a computer would remain unaffected.

Name _____

Date _____

Class _____

CHAPTER **24**

Electrical Remodeling

Complete the following questions and problems after carefully reading the corresponding textbook chapter.

_____ 1. Remodeling or modernizing an existing electrical installation is known as _____.

_____ 2. The _____ must be continuous from the service panel to all outlets.

3. Why should you always check for voltage even after turning off a breaker or opening a circuit?

_____

_____

_____

4. List five special tools that are needed for old work.

_____

_____

_____

_____ 5. *True or False?* Conduit is often used in old work, especially in finished walls.

_____ 6. When making wall openings for switch boxes, outlet boxes, and holes for pulling wire, the openings should be as _____ (large, small) as possible.

_____ 7. _____ is used to pull wires through bored holes and wall spaces.

_____ 8. *True or False?* New wiring can be concealed behind the baseboard.

_____ 9. Before working on an outlet, check it with a(n) _____ to make sure that the power is off.

10. List four ways to route wiring during remodeling.

_____

_____

_____

_____ 11. A box must be prepared and the cable attached to it _____ (before, after) the box is installed.

_____ 12. When modernizing the service of a building, most often, electrical service equipment with a higher _____ is the best solution.

_____ 13. *True or False?* When updating service equipment, make sure to provide a main service panel large enough for all the existing circuits, new circuits, and some spares for future expansion.

_____ 14. When installing a subpanel, it is usually mounted as near to the _____ as possible.

Describe the Code violations shown in the following diagrams.

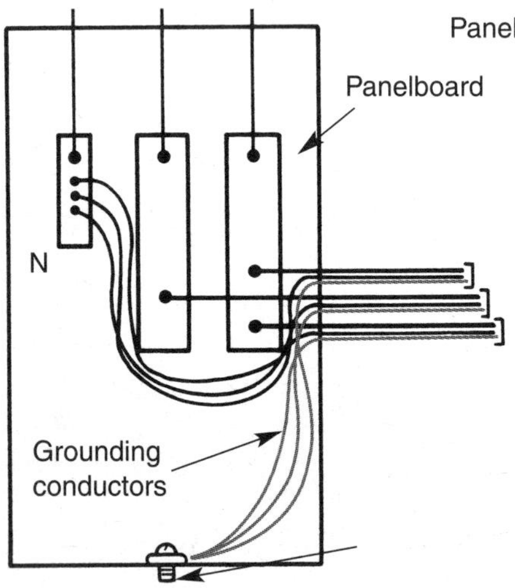

120/240 from service equipment

Panelboard

N

Grounding conductors

15.

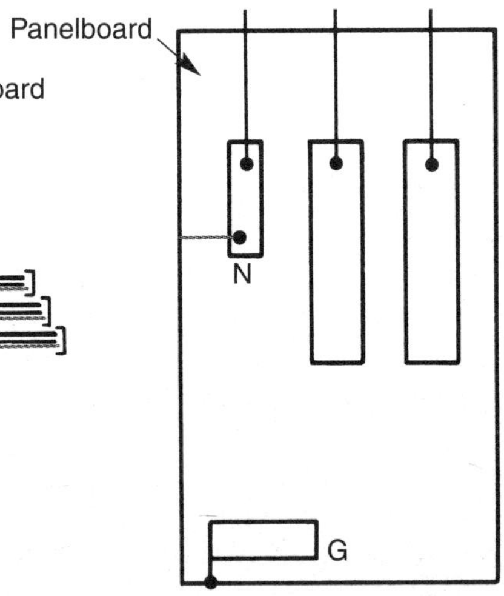

120/240 from service equipment

Panelboard

N

G

16.

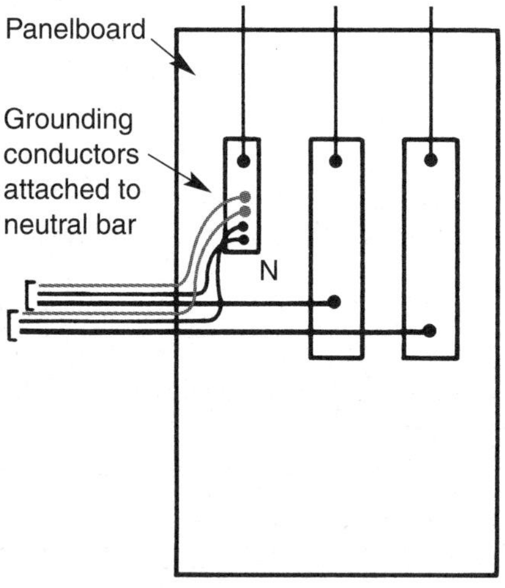

120/240 from service equipment

Panelboard

Grounding conductors attached to neutral bar

N

17.

Goodheart-Willcox Publisher

15. _____

16. _____

17. _____

Name _____

_____ 18. *True or False?* When grounding new service and subpanels, the grounding bar must be connected or bonded to the neutral bar.

19. What is the main benefit of a surface wiring system?

_____

_____

_____

_____ 20. *True or False?* When installing surface wiring, a typical installation begins with bringing the line side conductors into the base of the surface assembly.

NOTES

Name _____

Date _____

Class _____

CHAPTER **25**

Maintenance and Troubleshooting

Complete the following questions and problems after carefully reading the corresponding textbook chapter.

_____ 1. When troubleshooting, _____ must be uppermost in the electrician's mind.

_____ 2. Which tools are of particular importance when troubleshooting?
    A. Screwdrivers and pliers
    B. Drills and drill bits
    C. Meters and testers
    D. Crowbars and chisels

3. How do you determine if a receptacle is operating properly?

_____

_____

_____

_____

4. Explain how to check the ground continuity on an older-type receptacle that does not have a ground slot.

_____

_____

_____

_____

_____ 5. To test a switch, remove the cover plate and determine if there is power to the switch by touching one lead of the neon tester to the frame of the _____ and the other lead to the line side terminal.

_____ 6. *True or False?* It is often more convenient and economical to use one three-wire cable than two two-wire circuits.

7. What is the most common problem that occurs with split-wired or multiwired circuits?

_____

_____

_____

_____ 8. *True or False?* If a fuse blows, it should be replaced with one that has a higher rating.

9. List two ways in which you can identify a blown cartridge fuse.

_____

_____

_____

_____

_____ 10. If a breaker trips, you should ____.
    A. find the cause before resetting the breaker
    B. reset the breaker immediately
    C. never reset the breaker again
    D. None of the above.

_____ 11. *True or False?* All electrical problems are traceable through safe, careful analysis of the system components and perseverance on the part of the troubleshooting electrician.